50-50 everything means as I observe

I am.

Self Fulfilling Prophecy

Ilexa Yardley

=

Philosophy, physics and psychology (attempt to) articulate the human relationship to nature and also the underlying (overriding) nature of nature.

Thus, the quest for a theory-of-everything, unification, integration, individuation, self-actualization, absolute relativization, is the 'true' objective of science and also the 'true' nature of nature.

That is, the basic archetype of nature is the circular-linear nature (cyclical archetype, symbolic relationship) within nature.

The observation 'nature' is the core dynamic (in) (as) 'nature' means 50-

50 is the standard deviation and the self-fulfilling prophesy called 'nature...'

Form and substance share a circle, thus the essence of nature is duplicitous. The singular body called 'nature' is, by nature, ambivalent (duplicitous, ambiguous). Thus, the self-fulfilling prophecy.

Also articulated 'singular body problem' is the same as 'the zero body problem...' Thus, the duplicate body problem produces self-fulfilling prophecies.

Thus, observing nature as a 50-50 two state system, by observing the hidden symbolic circle between any X and Y (X and X, X and X') produces unification for all systems (symbolic, real) and, thus, the 'true' nature of nature.

Singularity in nature is duplicitous.

Thus, the 50-50 standard deviation is the self-fulfilling prophecy and, also the essence of nature. Thus ambiguity.   Ambivalence. Observation errors. The unit is duplicitous.

Thus a circular theory, using circular logic, is necessary, to adequately define and, thus, understand, the self-fulfilling prophecy of nature.

In other words, the cycle within nature is a singular deviation, thus, 50-50 is the norm. Thus, the ambiguous nature of nature. Ambiguity is nature. Observation errors force nature. Thus the self-fulfilling prophecy: nature.

Thus, a duplicitous unit (symbolizes nature).

Thus zero and one is the curve in a circular relationship with the line because the line and circle represent symbolic (1) in a circular relationship with real (0).

Thus the essence of the cycle is a circle. A duplicitous unit. Meaning, observation is necessarily erroneous. Thus the self-fulfilling prophecy.

Double is an alternate expression for single, which is an alternate expression for zero which is an alternate expression for triple.

Therefore, ambivalence (ambiguity), universally, (the object) (subject) is real. In numeric terms, 50-50. Forcing: observation. Errors. Self-fulfilling prophecy.

Thus life is the defense mechanism for death (and vice versa). Thus life is, again, by nature, ambiguous. Ambivalent. A duplicitous unit.

Thus observer-observation share a 50-50 circle. Thus whatever the observer observes is 'true...' Thus the alternate (observation) is, also, 'true...'

Thus the self-fulfilling prophecy.

Relativity

As a physicist, Albert Einstein identified the relative nature of nature as the singular body within nature. Everything is relative by observation and deduction. Thus, energy, mass, light are the three bodies in nature comprising the singular body called 'nature...' [1]

As a psychologist, Carl Jung identified the archetypical nature of nature. Everything is archetypical by observation and deduction. Thus individuation, self-actualization, integration are the three bodies in nature comprising the singular body called 'nature...' [2]

As a mathematician (philosopher) Emmy Noether identified the conservational nature of nature.

Everything is conservational by observation and deduction. Thus conservation, symmetry, unification, by conservation, observation and deduction, are the three bodies in nature comprising the singular body called 'nature...' [3]

Thus, integrating the three archetypes of nature (relativity, archetype, conservation) provides the correct explanation and the essence of nature. Life and death articulate a circle. 50-50. Life is a defense mechanism for death.

Thus pi is the constant coupling constant.

Thus ambiguity controls reality. Ambiguity is reality. Thus ambivalence. Duplicity. A duplicitous unit. 50-50. Observation. Errors. Self-fulfilling prophecy.

That is, individual and individual, individual and group, share a relative, archetypical, conservational circle.

The conservation of a circle is the reason for an object. Object is an alternate symbolic expression for a unit. Thus the essence of nature, object, unit, is a circle (also known as cycle). 50-50. Self. Fulfillment. Prophecy.

That is, Einstein, Jung and Noether advocated (articulated) the integration of opposites (yin-yang) as the core dynamic in nature.

Yang and yin are male and female are line and circle because the line is conserved by the circle and the circle is conserved by the line.

Conservation of the circle, then, is the basis for a unit, also known as object.

Thus both unit, and object, are fungible. Thus, sex. Death. Ambiguity. Ambivalence. Duplicity. Observation. Errors. Self-fulfilling prophecies.

Thus, the integration of physics, psychology and mathematics (and also philosophy) then, provides the insight needed to understand differentiation, where individuation and differentiation are a relative, yin-yang, complementary, oppositional cyclical, unifying archetype in nature.

Thus, the essence of any object in nature is the essence of a symbol.

Father-mother-child is parent-child. Yes is no. Here is go. Life is yes. Death is no.

Ambiguity. Ambivalence. Duplicity.

Observation. Errors. Self-fulfilling prophecy.

Any singularity is duplicitous (the single body always shares a circle with some other single body). Thus, again, ambiguity is real. Duplicity. Ambivalence. 50-50.

Zero shares a circle with one because the line is both diameter and circumference of a circle.

Things break apart. Things come together. Creation and destruction share an archetypical circle (more normally thought of as a cycle). Also known as object. More correctly designated: circle.

More realistically: sex in a circle with death as a circular metaphor for reality. Thus zero-one is one-zero. Defining ambiguity. Either-or. Both-and. Neither-either.

The duplicitous unit. The duplicity of a unit. 50-50. Observation. Errors. Self-fulfilling prophecy.

Background and foreground, any system, any entity, any process, share a mandatory circle. What makes things 'different?' is answered by 'what makes things 'alike...'".

This observation (deduction) (archetype) has broad and deep implications for humanity. However, you will have to 'think it through...' to get there. (Observation: error).

Self-fulfillment. Prophecy.

Conservation

The individual shares a circle with the group, any system, symbolic, real.

This means: the zero shares a circle with the infinite. Thus, the duplicate-body problem is the single body solution and the single-body problem is the duplicate body solution. Where zero and one are one and two.

Thus three and four are two and three are one and two are zero and one.

Thus five articulates one-half (two) in base ten, 101 in binary.

Thus 10-01. Where one-zero is hidden. Ambiguous. Duplicitous. 50-50.

Thus background shares a circle with foreground, thus single body system is a circle (cycle, circuit, switch).

Therefore, any object is a circle. Also light cone, hourglass, chalice. Clock, compass, lighthouse. Mirror. Shadow. Kaleidoscope. Brain. Stomach. Reproductive system(s). Observation. Error. Self-fulfilling prophecy.

That is, the movement of least resistance, reversion to the mean, is always diameter of a circle, meaning, two (not-one) is the basis for mathematics. X is not possible without X (X, X').

Two (not-one), then, is, also, the basis for linguistics. And two (not-one), then, is, also, the basis for symbolic systems in general, universally. 50-50.

Thus the essence of reality shares a circle with a symbol. Thus the basic symbol: life and death articulate a circle. 50-50 universally.

Thus life is protection (defense mechanism) (projection) for death and death is protection (defense mechanism) (projection) for life.

That is, a symbolic system shares a mandatory circle with a real system, thus the essence of a universal system. Thus the basis for statistics. And objects (objectivity). Ambiguity.

Observation. Errors. Self-fulfilling prophecy.

Repetition articulates a line, diameter of a circle. Thus two (not-one) is the basis for mathematics (any symbolic system, any physical system). Ambivalence. Duplicity.

Thus complementary opposition is real (a unit hides its opposite). Thus ambiguity (is real). Constant coupling. Coupling constant. Pi. 50-50.

Thus, any observation 50-50 is eventually true. Thus right-wrong, good-bad, true-false. Observation. Errors.

Self-fulfilling prophecy.

Archetype

Carl Jung got into trouble with his mentor, as everybody knows, because he refused to jettison his beliefs in astrology. His archetypical view of reality is based on his fundamental awareness (and knowledge) of astrology.

He used many different archetypes to articulate his fundamental knowledge about and belief in astrology, integrating it with psychology.

These include: dream, active imagination, individuation, mandala, conscious, unconscious, collective unconscious, extrovert, introvert, complex, synchronicity, unus mundus, szyzygy, integration, opposite. Intuition. Action. Love, hate. Life, death. Shadow. Active. Passive. Behavior. Anima. Animus.

Thus we begin by noticing (observing) astrology as the grandmother of all science and religion, mathematics and psychology. Also, economics and politics. History.

Astrology was (and still is) an ancient way to articulate man's relationship to nature, man's relationship to light, man's relationship to himself, man's relationship to others.

Fire, earth, air, water, represent one, two, three, four on a circle (also known as cycle, circuit, switch).

Light-dark. Heaven-earth. Active-passive.

Often, clock, hourglass, chalice. Compass. Microscope. Telescope.

The essence is ambiguous (a circle). Thus, 50-50. Also, and, alas, the many-body problem.

Thus, the zero-body problem.

Therefore the object is a subject. Man is always reflected by his relationship to the sun. 50-50.

Therefore man is always object to the sun. The essence of the sun (ambiguous). Light. Dark. Heaven. Earth.

Thus, the duplicity of a unit (the circular relationship between any object and the sun).

Thus any object shares a circular relationship with its opposite. Thus, the shadow. Also, cardinal, ordinal, fixed. Thus, mutable. Movement in a constant circular relationship with stillness (constant, variable).

So, thus...

Fast-forward all the way to Einstein, and it is common knowledge, now, the speed of light, as a constant, joins and separates matter and light as a relativity (most often designated: matter and energy as a constant).

Thus, we see, light integrates Einstein, Jung and Noether. Also, mankind, and nature. Also, then, all of the objects in nature. All of the essences in nature. All of the ambiguities: nature. All of the duplicities: nature. 50-50.

Thus the integration of the circle as a point, and, thus, the line, is the basis for one-two-three, one-two-one, zero-one, one-two, two-three, three-four. Duplicity. Unit. Light.

Thus object (in a circle with, and also named, then, subject). Thus essence (form and substance, real and symbolic). Conservation of the circle is the basis for a unit, symbolic, real.

Conservation of the circle is identified: duplicitous unit. 50-50.

Light: Constant, Variable

So we articulate light as the conservation of two constants, and this provides the understanding necessary to understand the essence of nature providing man with an answer for the origination (and destination) of any object in nature.

Where constant is an alternate name for variable. Thus, object is real. The unit is a circle (in a circle with the circle, as a circle within a circle).

Thus X and X (constant, variable) are necessarily a circle. The line is always diameter, and thus circumference, of a circle. This provides the understanding: man is an object and light is an object. Any symbol is an object.

Thus light is not moving, and, thus, everything moves.

Thus the essence of light (and man) is movement (in a circular relationship with stillness). Light and dark articulate a circle. Thus neither moves (so both can appear to move). 50-50.

A single body is, most basically, a zero body, meaning, it is not possible to have a single without some other single (also known as zero, infinity).

Thus the essence of a man is duplicitous. Thus the essence of nature is 50-50. Ambiguous. An empty tent (potential movement, mind and matter as a circle). The chalice and its host.

Thus, hydrogen, unlike all the other elements, is hiding its neutron within (as) all the other elements. Thus hydrogen, when burned, produces water.

Hydrogen and water (also known as oxygen) share the ever-present (never-present) circle. A standard deviation. Thus its essence: duplicitous.

Thus carbon is an alternate expression for hydrogen is an alternate expression for oxygen because the single body is necessarily duplicitous. The line is always diameter and circumference of a circle. Thus its essence: duplicitous.

A single body must reproduce a single body. Thus the single body is, necessarily, a duplicitous body (individual shares a constant circle with a group). Thus its essence: duplicitous.

Thus hydrogen is a basic binary system around which all the other systems rotate and revolve explaining, first and foremost, the basic relationship between time and space (light and sound) (one body and another). Thus its essence: duplicitous.

The reproduction (conservation) of a circle demonstrates the reproduction (conservation) of hydrogen (water). Thus hydrogen and oxygen (any two elements) articulates a circle (one and two, one as two).

An element is an object. Thus, essence is duplicitous. Thus, shadow. Sex. Death. Ambiguity (duplicity, the lie). The zero-one relationship. An empty tent. The unit is (always) duplicitous. Thus, 50-50 is the basis.

The duplicitous unit, then, ignores the symbolic representation of one unit, assigns it to the symbolic representation of two units, using the assumption, line is both diameter and circumference of a circle. Duplicity defines reality (mind is matter).

Zero shares a constant (variable) circle with its one.

Thus good and bad behavior (right and wrong, yes and no, true and false). Life and death articulate a circle. Thus zero, one. 50-50.

Symbols

In scientific method, we identify the ingredients and the process as an input and we measure the output as a result.

Thus in nature, to understand the core dynamic in nature, we identify the ingredients as a basic line and circle, we identify the process as observing line and circle between and, thus, around, any X and Y (X and X, X and X') and thus we have our result (in a circle, thus a cyclical relationship, with its cause).

Thus the basic cycle is the core dynamic in nature. Symbolically aligned. Thus subject-object-subject is, always, subject-object (any combination).

This is confusing and, at first, heretical.

We have taken up the jettisoned circular logic which was forbidden sometime way far back in time. That is, the jettison, then, depends on a preliminary assumption, 'far back,' 'in,' 'time...' Thus, here, we show, and we can see, all of us, every observation involves a preliminary assumption.

50-50 is, thus, the most basic assumption. Thus, the archetype is a basic objectivity describing behavior. Any symbol is an object (describing behavior).

That is, nature, by observation, assumes nature (by deduction)...

Thus to observe nature is real is, also, to observe nature is symbolic. Thus essence is symbolic in order for essence to be real.

This is cyclical and circular because observation and deduction share a circle (thus all observation is 50 percent flawed). Observation, then, is the core dynamic (the symbolic reality) in (of) nature.

Therefore one-half is one-third. 50-50 is 50-50-50. Diameter is circumference. Thus the observation 'single body' is, necessarily, flawed.

Observation is dependent on a 50-50 circle.

This is also designated two-state-system, three-state-system (where two and three, like one and two, and zero and one, are obviously, a circle).

Thus the essence of one is two and the essence of two is one. A two-state system is real (single-state).

Where state and body are the same (subject and object share a mandatory circle).

The observation, 50 percent, assumes a decimal reality, and here we can discern where all observational mistakes are made, why and how they are made, and how they can be used to advantage (to provide the 'other side' with a disadvantage).

Binary is decimal (zero and one sharing a circle). Therefore five is three is two. Therefore 50-50 is 50-50-50 (horizontal is vertical).

100 is 10, 01. 10, 00. Thus, you are continually entangled with your 'opposite...'

Thus three chambers of death: nervous system, reproductive system, digestive system. Complementarity. Uncertainty. Entanglement All of these, ambiguous. 101, 10, 01.

Thus, to discern the natural order of nature, we begin by observing the natural chaos of nature.

Therefore, chaos and order share a very simple circle where you cannot have one without the other. That is, chaos is a deduced assumption called logic. Thus a basic (standard) deviation.

Or, let's put it another way, order is a deduced chaos called logic.

Whatever. You are not confused. Thus, you are, confused. That is negation and duplication are the same. Negation, is, thus, affirmation.

Any symbolic expression articulates movement of a line, thus begin and end articulate a circle, because the line, of any length, is always diameter, and thus, circumference, of a circle.

Thus the essence of a unit is a circle.

So draw a circle and then ask yourself... have I drawn circumference or diameter? You will notice, easily, you have to have one in order to have the other.

So either observation is true, and false, or false, is false. Now we find the order we all 'desperately' seek.

A singular subject is a duplicitous object. A singular object is a duplicitous subject. These share a circular relationship.

Thus the essence of either determines the essence of both where essence is an alternate word for archetype, an alternate word for circular relationship between real and symbolic.

Thus an array of ones is equal to an array of zeroes thus infinity is real.

And thus the symbol articulates reality. Point line and circle are pi diameter circumference. Thus the symbolic reality: ambiguity (two units articulate(s) nature) (joined and separated by a circle).

So, there it is. It's simple.

Any X and Y articulate a circle. This is because the line is always diameter of circle. The line is thus circumference of a circle. Thus, it's all 50-50 (so half will have to say... it's never 50-50...). Thus 50-50-50 is 50-50.

So, obviously, now you know yourself... you cannot have one without the other, and this is the easily deduced and thus observed core dynamic (essence) of nature.

For some, symbolically, thus real: a duplicitous unit.

Reality

Input-process-output (one-two-
three) is input-output-process
because the circle (cycle) between
input and process and process and
output unify input, output, process.

Therefore, the nature of nature is
cyclical (reproduction is observation
and observation is reproduction)
producing self-reference. Nervous
system. Reproductive system.
Digestive system. Observation, any
system.

That is, me-not-me is you-not-you.
They share a circular relationship.

Thus the circle is the object (and
also it's the subject) beneath and
over nature. This explains the core
dynamic: conservation of the circle

(circles) as the nature (core dynamic, essence) of nature.

Therefore, it is easy to deduce, the circle is the essence in nature, thus the singularity (duplicity) (triplicity) called 'nature...'

Therefore, 50-50 is the base for, and thus the basis of, and thus the essence of nature.

An object (objectivity) is movement in a circular relationship with stillness (time and no-time, time and all-time, zero and infinity, essence). Ambiguity. Duplicity. Unity.

The line is always diameter of a circle.

Thus 50-50. Nature. Unit.

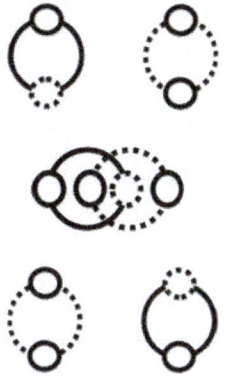

Ambiguous Representation Basis for
Universal Symbolic System

What this article says (in plain
English, conventional ordering):
there is a hidden circle between any
X and Y (X and X', X and X) because
the line between X and Y is diameter
and circumference of a circle.

Thus, two (not-one) is the foundation (and the correct name) for a unit (any unit). Thus, from the (hidden, abstract) circle's point of view all units are the same (joined and separated by a circle).

Thus 'circle' is the correct observation for a noun and verb (a circle can only circle) and 'circle' is the correct observation for any unit (object, subject). Unit (object) is fungible.

Thus we see astrology in man's earliest rendering of nature (symbolic representations of constellations on cave walls). Thus astrology formed the foundation for mathematics (me-not-me, you-not-you, it-not-it).

Symbolic representation shares a circle with physical actuality, thus circle (cycle, circuit, switch) (hourglass, light-cone, chalice). Lighthouse. Mirror. Kaleidoscope. Background. Foreground. Host.

Astrology, then, also formed the foundation for archetypes (symbolic behavior found in religious, later, Greek plays, mythological metaphors). Thus astrology found its way into psychology (Jung's archetypes), and thus psychology finds its way into physics (individuation, self-actualization, identity).

Thus, we have a nice, neat circle (mathematics-philosophy, psychology-physics).

Notice thus, a circle between and around mathematics-philosophy, psychology-physics, the individual letters that make up words that make up sentences and paragraphs, thus, ideas.

Ideal shares a circle with real.

Thus, conservation of a circle is a symbolic representation of cyclical reproduction in nature (reproduction in general). Thus an object shares a circle with a subject (some other object). One unit begets two units and vice versa.

Thus three units is the repetition of two units is the repetition of one unit. Therefore, two (not-one) is unit (basic objectivity, 50-50). Thus, astrology forms the foundation for mathematics (and, also, movement).

Three is not possible without two is not possible without one. The circle is the basis for two units.

Thus, we cannot move ahead once (unless) we recognize (realize, observe, understand) the underlying (overriding) circle (also known as cycle) (circuit, switch).

Thus, pi is easily the only observer in nature (also known as 'mind...').

Thus, a unit, is dependent on an anti-unit (one, two) and negation is duplication. (And, or, is, is-not).

Ambiguity is ambivalence. Unit is and is-not anti-unit. Is and is-not are the same. Negation is, thus, affirmation. Duplication is creation and, also, destruction because creation and destruction, like any X and Y, share an uber-basic circle.

Thus X, XX, XXX is the same as X is XX is XXX is the same as X is-not XX is-not XXX (from the circle's point of view, from pi's point of view, from mind's point of view, from God's point of view) (where circle, pi, mind, God are archetypes for unit).

The archetype one shares the archetype circle with the archetype two. This is a basic syzygy.

Thus one-two-three is beginning-middle-end is input-process-output is a three-state-system (symbolic, real) is a two-state system (real, symbolic).

Where the 'state'ment 'line is diameter and circumference of circle' articulates and demonstrates the origination and, also, the destination of 'state' (physical, psychological, philosophical).

Thus zero and one (0-1, 1-0) is the informational (symbolic) and, also, the physical (real) state.

Where zero and one share a circumference and diameter. Thus essence (circularity) ambiguity (duplicity) is real. Thus the duplicitous unit (is ambivalent),

Thus single body system is zero body system is multiple body system is the n-body system, where two (not-one) is the natural limit in all directions (mediated by pi).

Thus an array of zeroes collapses as an array of ones expands, and a singular zero is the same as (from the circle's point of view, from pi's point of view, from 'mind's' point of view) a duplicitous 'one....' (two units, not-one).

Thus, basic objectivity (50-50).

The circle is the basic archetype for unit, thus the absolute cycle in nature, thus, the nature of nature is dependent on the conservation of a circle (reproduction in general). M,F is 0,1 is F,M is 1,0.

Thus the circle is the (constant) basis for a unit (variable) (object).

This analysis provides the unification (singularity) humans seek to understand (enjoy, utilize).

Thus, unification (duplicity) is the nature of nature. Two subjects, one object. One object, two subjects. Also known as...

Conservation of the circle is the basis for... what is also known as... 50-50 ... Observation.... Error(s)... Self-fulfilling... Prophecy...

Summary

Every unit (from the circle's point of view) is a circle thus pi is the dimensionless coupling constant in the background thus 50-50 everything in the foreground. This lets all of us breathe easier. Meaning we expect the 50-50, plan for 50-50-50 and thus it's 50-50-50-50 (taking the other half into account).

Thus we're intelligently prepared, (infinitely aware) and completely objective. We understand where objectivity originates. It is: our essence. The duplicitous unit. The universal metaphor: balance in nature. Man's symbolic relationship to nature, symbolic behavior universally, is conservation (thus reproduction) of a circle.

Any movement makes the opposing movement real.

π

(1) <u>Albert Einstein</u> (1905) "<u>Zur Elektrodynamik bewegter Körper</u>", Annalen der Physik 17: 891

(2) Jung, C. G. (1934–1954), The Archetypes and The Collective Unconscious, Collected Works 9 (1) (2 ed.), Princeton, NJ: Bollingen (published 1981), <u>ISBN</u> <u>0-691-01833-2</u>

(3) Emmy Noether; Mort Tavel (translator) (1971). "Invariant Variation Problems". Transport Theory and Statistical Physics 1 (3): 186–207. <u>arXiv:physics/0503066</u>. <u>Bibcode:1971TTSP....1..186N</u>. <u>doi:10.1080/00411457108231446</u>. (Original in Gott. Nachr. 1918:235-257)

CPSIA information can be obtained at www.ICGtesting.com
Printed in the USA
BVOW06s2225180316

440939BV00026B/226/P